Edexcel GCSE 9-1
Mathematics Practice Papers

Foundation

Copyright © 2017 Mark Ritchings

Paper 1
You must not use a calculator.

1) Write $\frac{7}{8}$ as a decimal.

$$8 \overline{)7.0^60^40}$$ = 0.875

..........0.875..........(1)

2) (a) Simplify $2 \times x \times 3 \times y$

$2 \times 3 = 6$
$= 6xy$

..........$6xy$..........(1)

(b) Simplify $12a + 3b - 4a + 6b$

$12a + 4a - 3b + 6b$
$16a - 9b$

..........$16a - 9b$..........(1)

3) Solve $\frac{y}{2} = 2.3$

$\begin{array}{r} 2.3 \\ \times 2 \\ \hline 4.6 \end{array}$

..........4.6..........(1)

4) Write 42.12374 correct to 3 decimal places.

42.132 [74] → bigger than 50
So the answer 42.133

..........42.133..........(1)

5) Work out the value of 3^4.

$3 \times 3 \times 3 \times 3$

$\begin{array}{r} ^2 27 \\ \times 3 \\ \hline 81 \end{array}$

..........81..........(1)

6) Work out 70% of 80.

$$\frac{70}{100} \times 80 = 56\%$$

...................(2)

7) (a) Work out $\frac{3}{5} \times \frac{7}{10}$

$$\frac{3 \wedge 2}{5_{\wedge 2}} \times \frac{7}{10} = \frac{6}{10} \times \frac{7}{10} = \frac{42}{10} \text{ or } 4.2$$

$\frac{42}{10} = 4.2$

...................(1)

(b) Work out $\frac{2}{5} + \frac{1}{3}$

$$\frac{2^{\times 3}}{5_{\times 3}} + \frac{1^{\times 5}}{3_{\times 5}} \qquad \frac{6}{15} + \frac{5}{15} = \frac{11}{15}$$

$\frac{11}{15}$

...................(2)

8) Bob buys

    3 notebooks costing £1.20 for each notebook
    1 pack of pens costing £1.60
    2 calculators

Bob pays with a £20 note. He gets £1.70 change. Work out the price of one calculator.

1.20
   3
―――
3.60
+1.60
―――
(5.20)

20.20
− 5.20
―――
=15.20

7.60
−1.70
―――
0.90

7.60
+ 90
―――
=£8.70 p

2)15.20
  7.60

...................(3)

9) Bob spins a fair, 5-sided spinner numbered 1 to 5.

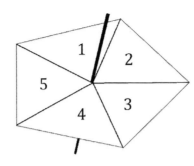

(a) Write down the probability that the spinner lands on 3.

................$\frac{1}{5}$................(1)

(b) Write down the probability that the spinner lands on 7.

................$\frac{0}{5}$................(1)

10) Express 140 as the product of its prime factors.

................................(2)

11) Work out 23.7 × 4.5.

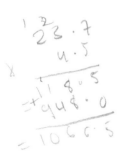

................1066.5................(3)

12) Complete the table.

Seconds	Minutes
40	$\frac{2}{3}$
24	
	$3\frac{1}{2}$

(2)

13) Write 270 m as a fraction of 3 km.

..........................(2)

14) There are five red beads, six green beads and nine blue beads in a box. There are no other beads in the box.
One bead is taken, at random, from the box.

Write down the probability that this bead is green.

..........................(2)

15) A sequence of patterns is made using white and grey square tiles.

(a) How many white tiles are needed to make the fifth pattern?

..........6..........(1)

(b) How many **tiles** are needed to make the eighth pattern?

..........(1)

Another sequence of patterns is also made using grey and white square tiles.

(c) How many grey tiles are needed to make the tenth pattern in this sequence?

..........(1)

(d) How many white tiles are needed to make the sixth pattern?

..........(2)

16) Bob owns some books. The number of each type of book that he owns is shown in the table below.

Type of book	Number of books
Manga	30
Maths	20
Science fiction	22

Draw an accurate pie chart to show this information.

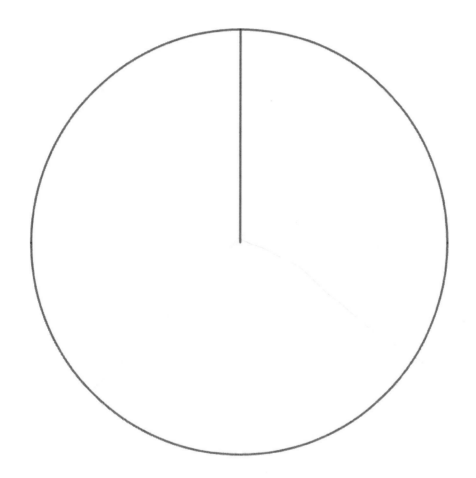

(3)

17) The diagram below show a trapezium and a rectangle. The area of the rectangle is three times the area of the trapezium.

Work out the length of the rectangle.

Diagram not drawn accurately

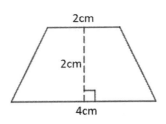

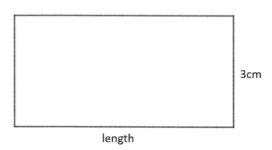

length

$$2 \times 2 \times 4 = 16$$

$$\begin{array}{r} 16 \\ \times\ 3 \\ \hline 48 \end{array}$$

$$\begin{array}{r} 16 \\ 3\overline{)4^18} \end{array}$$

.........~~48~~ 16.........(4)

18) Solve the simultaneous equations

$$3x + 5y = 13$$
$$15x = 90$$

$15x = 90$

$\dfrac{90}{15}$ $15\overline{)90}\ ^6$

So $15 \times 6 = 90$
$x = 6$

$3x + 5y + 15x = 13 + 90$
$3x + 15x + 5y = 103$
$18x + 5y = 103$

$3 \times 6 + 5y = 13$

$18 + 5y = 13$

$5y = 18 - 13$

$y = 5$

x =..........................

y =..........................

(3)

19) Calculate the length of AB in the diagram below.

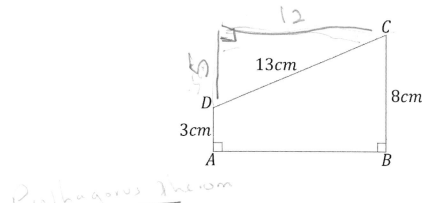

..........12..........(4)

20) Write down the equation of a line that is parallel to the line with equation $x + y + 1 = 0$.

..........................(2)

21) $s = ut + \dfrac{at^2}{2}$

$u = \dfrac{1}{2}$ $t = 2$ $a = 3$

Work out the value of s.

..........................(3)

22) (a) n is an integer.

$-4 < 2n \leq 6$.

Write down all the possible values of n.

....(3)

(b) Solve $6(x + 3) = 39$

$6(x+3) = 39$ $6x = 39 - 18$
$6x + 18 = 39$ $6x = 21$
$24x = 39$
                        $x = \frac{21}{6}$
$39/24 = 15$
                        $x = 3.5$

..........3.5..........(2)

23) The normal price of a certain guitar is £3500.
In a sale the price is reduced by 15%.
Work out the price of the guitar in the sale.

$\frac{3500 \times 15}{100} = 525$ So $3500 - 525 = 2975$

7500   10% 750
               275
          15% 1125

..........£2975..........(3)

24) 60% of the visitors to a school's summer fair are children. Find the ratio of the number of adults to the number of children. Write your answer in its simplest form.

..................(2)

25) ABC is a triangle. $\overrightarrow{CA} = \boldsymbol{a}$ and $\overrightarrow{CB} = \boldsymbol{b}$.

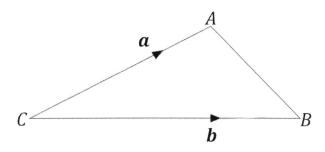

(a) Write down in terms of $\boldsymbol{a}$ and $\boldsymbol{b}$ the vector $\overrightarrow{AB}$.

..........................(1)

(b) Work out the vector $\overrightarrow{CA} + \overrightarrow{AB} + \overrightarrow{BC}$.

..........................(1)

26) The diagram below shows a regular octagon and an equilateral triangle.
Work out the size of the angle marked x.

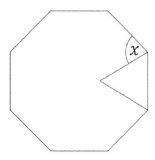

..........................(3)

27)

(a) Expand and simplify $(b+4)^2$

..........................(2)

(b) Expand and simplify $(c+7)(c-7)$

..........................(2)

28) Calculate the area of the triangle shown below.

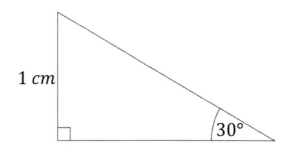

..........................(3)

29)
   (a) Factorise $x^2 - 2x - 15$.

   ....(2)

   (b) Solve $x^2 - 2x - 15 = 0$

   ....(1)

30) The shape below is made up of two parallel lines and two semi-circles. Calculate the shaded area.   Give your answer in terms of π.

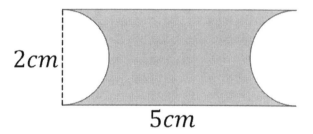

..........................(4)

Paper 2
You may use a calculator.

1) Using your calculator, or otherwise, work out $3\frac{1}{2} \div 2\frac{3}{4}$

   Give your answer as a mixed number in its simplest form.

   ....(2)

2) Write these numbers in order of size. Start with the smallest number.

   $\frac{3}{7}$ 0.4   43% $\frac{21}{50}$ 0.401

   ....(2)

3) Write down the coordinates of the point marked with a cross.

   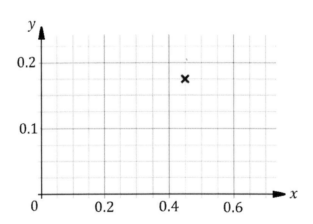

   ....(2)

4) Round 382.7 correct to one significant figure.

   ....(1)

5) Expand and simplify
$$2x(3x+5) - 2x(x-7)$$

...........................(2)

6) Find the highest common factor and lowest common multiple of 420 and 90.

....(4)

7) Here is a list of numbers.

      7   12   6   10   4   15   8   11

Find the median.

...........................(2)

8) The probability that Bob arrives at work late on any given day is 0.02. What is the probability that Bob is not late?

...........................(1)

9) Factorise $15x^2y - 3xy^2$

...........................(2)

10) $x = 13.6$ correct to one decimal place.
Write down the error interval for x.

..................................(1)

11) (a) Write down the first five prime numbers

..................................(2)

(b) Write down the first three cube numbers.

..................................(2)

12) Find the value of $\frac{(7.23-12.49)^2}{\sqrt{2.32}+4.656}$
Write down all the figures on your calculator display.

..................................(2)

13) (a) Write 6.31×10^5 as an ordinary number.

..................................(1)

(b) Work out the value of $(3.5 \times 10^8) \div (7 \times 10^2)$

Give your answer in standard form.

..................................(2)

14) (a) Describe fully the single transformation that maps shape A onto shape B.

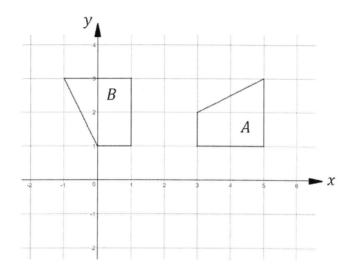

....(2)

(b) Reflect triangle C in the line $x = 1$.

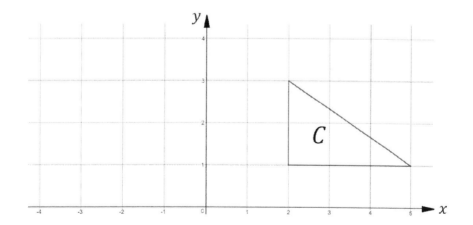

(2)

15) A person's body mass index is calculated using the following formula,

$$BMI = \frac{m}{h^2}$$

where m is the person's mass in kilograms and h is the person's height in metres.

1 foot = 12 inches

1 inch = 2.54 cm

1 kilogram ≈ 2.2 pounds

1 stone = 14 pounds.

Bob weighs 12 stone 5 pounds. Bob's height is 6 feet 2 inches.

By first calculating Bob's weight in kilograms and his height in metres show that Bob's BMI is 22.3 correct to one decimal place.

..........................(5)

16) After a 15% price increase a ticket costs £74.75.
What was the price of the ticket before the price increase?

..........................(3)

17) The diagram show a quadrilateral, $ABCD$.
Work out the value of x.

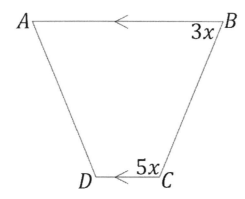

..........................(4)

18) The heights of 32 people are summarised in the table below.

Height, h (metres)	Frequency	
$1.5 \leq h < 1.6$	3	
$1.6 \leq h < 1.7$	5	
$1.7 \leq h < 1.8$	10	
$1.8 \leq h < 1.9$	12	
$1.9 \leq h < 2.0$	2	

(a) Write down the modal class interval.

..........................(1)

(b) Calculate an estimate for the mean height of the 32 people.

..........................(3)

19) The nth term of a sequence is $3n^2 - 2$.
Work out the sixth term of the sequence.

..........................(2)

20) Here are the speeds of 12 cars, in km/h.

| 21 | 33 | 42 | 19 | 32 | 25 |
| 46 | 37 | 28 | 33 | 35 | 42 |

(a) Show this information in a stem and leaf diagram.

..........................(3)

(b) Work out the median speed.

..........................(1)

21) The diagram below is a 3 dimensional sketch of a prism.

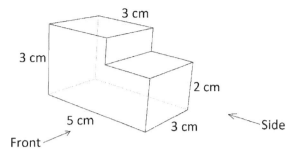

On the grid below draw accurately the front and side elevations of the prism.

(3)

22) You can use this graph to change pounds to kilograms.

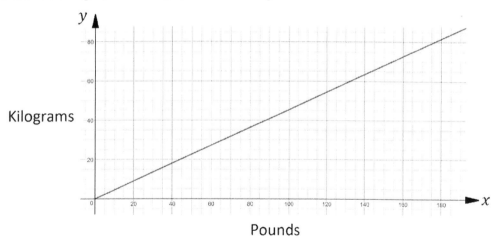

Change 155 pounds to kilograms.

..........................kilograms (1)

23) $v = u + at$

Make t the subject of this formula.

..........................(2)

24) Ann and Bob share some money in the ratio 7:4.
Ann receives £24 more than Bob.
How much money does Ann receive?

..........................(3)

25) The diagram below shows two right angled triangles.
$AD = 5$cm. $CD = 3$ cm. Calculate the length of BC.

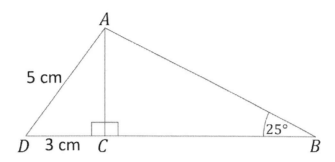

..........................(4)

26) Bob has to paint a wall. The wall is 6 m long and 2.5 m high.
One litre of the paint Bob will use covers an area of 5 m².
The paint is sold in tins containing 2 litres.
The paint costs £4.75 for one litre.
How much money will Bob need?

..........................(5)

27) Work out the size of the angle marked x in the diagram below.

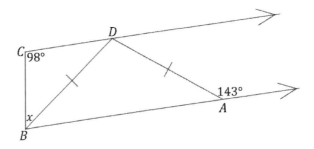

..........................(4)

28) A cylindrical container has a radius of 15 cm and a height of 40 cm.
Bob pours 15 litres of water into the container. How much more water could the container hold?

..........................(4)

Paper 3
You may use a calculator.

1) (a) Write the value of $\frac{11}{12}$ as a decimal, correct to three decimal places.

..........................(1)

(b) Work out $\frac{3}{5}$ of 120.

..........................(1)

2) (a) Write the following numbers in ascending order.

   0.301   0.3   0.03   0.031   0.31   0.003

....(1)

(b) Write the two numbers, from the list, which have a product of 0.009.

..........................(1)

3) (a) Three fifths of the students in a class are girls.
   What is the ratio of the numbers of girls to the number of boys.

..........................(1)

(b) The ratio of the number of male teachers to the number of female teachers, in the school, is 5:7.

If there are 36 teachers, how many of them are female?

..........................(2)

4) Bob thinks of a number.
   He divides the number by 5 and multiplies the result by 4 to get an answer of 28.
   What number did Bob think of?

..........................(2)

5) Large packets of exercise books contain 50 books.
   Small packets contain 20 books.
   Bob buys x large packets and y small packets.
   Write an algebraic expression for the total number of books that Bob buys.

..........................(2)

6) Amy and Bob share £96.
   Amy receives 7 times as much as Bob.
   How much money does Amy receive?

..........................(2)

7) Correct to two significant figures, there are 1200 students at a school.
   (a) What is the largest possible number of students?

..........................(1)

   (b) What is the smallest possible number of students?

..........................(1)

8) Bob sells a house for £191880 and makes 23% profit.
   How much did Bob pay for the house?

..........................(2)

9) Here is a list of numbers.

23   29   21   27   20   29   22   19

(a) Work out the mean of the set of numbers.

..........................(2)

(b) Work out the range of the set of numbers.

..........................(2)

(c) Write down the mode of the set of numbers.

..........................(2)

10) $v = 5 + 6t$

(a) Work out the value of v when $t = 7$.

..........................(2)

(b) Make t the subject of the formula.

..........................(2)

11) Bob has four tiles.

$$\boxed{2} \boxed{5} \boxed{6} \boxed{8}$$

Each tile has a number on it.
Work out how many different odd numbers Bob can make using all four tiles.

...............................(2)

12) Last year at Bob's school 240 students took GCSE maths.
$\frac{1}{20}$ got grade 9.

15% got grade 8

The ratio of grade 7 passes to grade 6 passes was 3:5.

No-one at the school got a grade lower than 6.

Work out the number of grade 6 passes at this school.

...............................(5)

13) (a) Write down the mathematical name of the shape shown below.

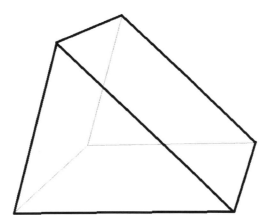

....(1)

(b) Write down the number of

Edges...........................(1)

Vertices...........................(1)

Faces...........................(1)

14) Bob bought a washing machine.
The washing machine cost £280.
Bob paid a 10% deposit and paid the rest of the cost in 12 equal monthly payments.
How much was each payment?

..............................(2)

15) Solve the simultaneous equations

$$5x - 4y = 5$$

$$5x + 2y = -7$$

..........................(3)

16) (a) Complete the table of values for $y = \frac{1}{2}x^2 - x$

x	-2	-1	0	1	2	3	4
y							

(2)

(b) On the grid below draw the graph of $y = \frac{1}{2}x^2 - x$ for values of x from -2 to 4.

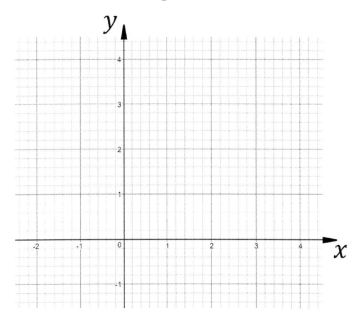

(2)

17) a and b are integers.
$ab = -12$
$a + b = 4$
$a > b$

Find the value of a and the value of b.

....(2)

18) The diagram below shows a prism.

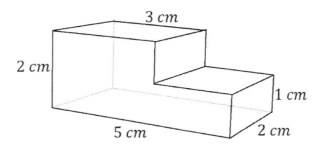

(a) Calculate the volume of the prism.

..............................(3)

(b) Calculate the surface area of the prism.

..............................(3)

19) (a) Solve the inequality $6x - 2 < 2x + 7$

...........................(2)

(b) Show the inequality $-3 \leq x < 2$ on the number line below.

...........................(2)

20) Bob drives 20 miles at an average speed of 40 mph.
He then drives a further 50 miles at a speed of 60 mph.
Work out Bob's average speed for the whole journey.

...........................(4)

21) Bob is planning to buy a bookcase for his study.
The side of the bookcase measures 28 cm by 2.02 m.
The ceiling in Bob's study is 204 cm above the floor.
Once assembled will Bob be able to stand the bookcase up?

..........................(4)

22) Work out $1.3 \times 10^7 \times 3.6 \times 10^{105}$
Give your answer in standard form.

..........................(2)

23) Expand and simplify $(2x + 3)(3x - 4)$

..........................(2)

24) Show that triangles ABC and DEF are congruent.

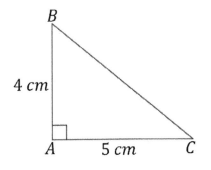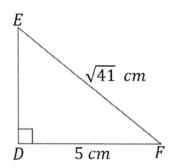

..........................(5)

25) Bob has 4 red cards and 4 black cards. The eight cards are numbered 1 to 8.
Bob selects a card at random.
Is it possible to calculate the probability that Bob selects an even numbered red card?
You must explain your answer.

..........................(4)

Paper 1 Answers

1) $\frac{7}{8} = 7 \div 8$.

$$8 \overline{)7.000}^{764} = 0.875$$

2) (a) $2 \times x \times 3 \times y = 2 \times 3 \times x \times y = 6xy$

   (b) $12a + 3b - 4a + 6b = 8a + 9b$

3) $\frac{y}{2} = 2.3$
   Multiply both sides by 2
   $y = 4.6$

   Alternative method

   $y \to \div 2 \to 2.3$

   $y \leftarrow \times 2 \leftarrow 2.3$

   $y = 2.3 \times 2$

   $y = 4.6$

4) 42.12374 is between 42.123 and 42.124. Since it is greater than 42.1235 it is closer to 42.124.

5) $3^4 = 3 \times 3 \times 3 \times 3 = 81$

6) 70% of 80 = $\frac{70 \times 80}{100} = \frac{5600}{100} = 56$

   You could also work out:
   10% of 80 = 8
   70% of 80 = 7 x 8 = 56

7) (a) $\frac{3}{5} \times \frac{7}{10} = \frac{21}{50}$

   (b) $\frac{2}{5} + \frac{1}{3} = \frac{6}{15} + \frac{5}{15} = \frac{11}{15}$

8) Bob spent £20.00-1.70 = £18.30
The notebooks and pens cost 3 x £1.20 + £1.60 = £3.60+1.60 = £5.20.
The calculators cost £18.30-£5.20 = £13.10.
One calculator costs £13.10 ÷ 2 = £6.55.

Use appropriate written methods for all of these calculations.

9) (a) $\frac{1}{5}$ (b) 0

10) You could use a factor tree:

```
       140
      /   \
     10    14
    /  \  /  \
   2   5  2   7
```

$140 = 2^2 \times 5 \times 7$

11) 23.7 x 4.5

x	200	30	7
40	8000	1200	280
5	1000	150	35

```
  237
 x 45
 1185
 9480
10665
```

23.7 x 4.5 = 106.65

12) 24 seconds = $\frac{24}{60}$ minutes = $\frac{2}{5}$ minutes

13) $3\frac{1}{2}$ minutes = $3\frac{1}{2}$ × 60 seconds = 3 × 60 + $\frac{1}{2}$ of 60 = 180 + 30 = 210 seconds.

14) 5+6+9=20. There are 20 beads in the box. 6 of the beads are green.
The probability of taking a green bead is $\frac{6}{20} = \frac{3}{10}$

15)

Pattern number	1	2	3	4	5	6	7	8
White tiles	2	3	4	5	6	7	8	9
All tiles	3	5	7	9	11	13	15	17

(a) The fifth pattern has 6 white tiles.
(b) The eighth pattern has 17 tiles.

(c)

Pattern number	1	2	3
Grey tiles	4	6	8

The nth term of the sequence is 2n+2 so the tenth term is 2 x 10 +2 = 22

(d)

Pattern number	1	2	3
White tiles	2	6	12

The white tiles form rectangles. The lengths and widths increase by 1.
2 = 1 x 2   6 = 2 x 3   12 = 3 x 4

In the sixth pattern there are 6 x 7 = 42 white tiles.

16)

Type of book	Number of books	Angle
Manga	30	5 x 30 = 150
Maths	20	5 x 20 = 100
Science fiction	22	5 x 22 = 110
Totals	72	5 x 72 = 360

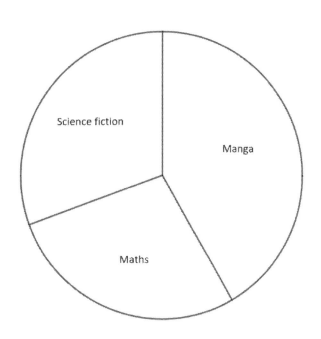

17) The area of a trapezium is half the sum of the parallel sides multiplied by the distance between them.

The area of this trapezium = $\frac{1}{2}(2+4) \times 2 = 6$ cm².
The area of the rectangle = 3 x 6 = 18 cm².

The length of the rectangle is 6cm.

18)
$15x = 90$
$x = \frac{90}{15} = 6$
Substituting into the other equation gives

$3 \times 6 + 5y = 13$
$18 + 5y = 13$

$y \rightarrow \times 5 \rightarrow +18 \rightarrow 13$
$y \leftarrow \div 5 \leftarrow -18 \leftarrow 13$

$13 - 18 = -5$
$-5 \div 5 = -1$

$y = -1$

$x = 6, \ y = -1$

19) Draw a line from D, parallel to AB.

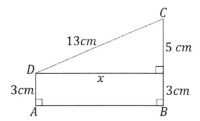

Then, by Pythagoras' theorem

$x^2 + 5^2 = 13^2$
$x^2 + 25 = 169$
$x^2 = 169 - 25$
$x^2 = 144$
$x = \sqrt{144}$
$x = 12 \ cm$
$AB = 12 \ cm$

20) The easy way to answer this question is to write $x + y + \text{anything} = 0$.
Alternatively you can rearrange the given equation to make y the subject.
$$y = -x - 1$$
Now change -1 to any other number to obtain $y = -x + 1$ for example.

21)
$$s = ut + \frac{at^2}{2}$$

$u = \frac{1}{2}$ $\qquad t = 2 \qquad a = 3$

$s = \frac{1}{2} \times 2 + \frac{3 \times 2^2}{2} = 1 + \frac{3 \times 4}{2} = 1 + \frac{12}{2} = 1 + 6 = 7$

22) (a) $-4 < 2n \leq 6$

Divide all parts of the inequality by 2 to get

$-2 < n \leq 3$

The possible values of n are -1,0,1,2,3

(b) $6(x + 3) = 39$

Expand the brackets	$6x + 18 = 39$
Subtract 18 from both sides	$6x = 21$
Divide both sides by 6	$x = \frac{21}{6}$
	$x = 3\frac{1}{2}$

Alternative method

$x \to +3 \to \times 6 \to 39$

$x \leftarrow -3 \leftarrow \div 6 \leftarrow 39$

$\frac{39}{6} = \frac{13}{2} = 6.5$

$6.5 - 3 = 3.5$

$x = 3.5$

23) 10% of £3500 = £350
    5% of £3500 = £175
    15% of £3500 = £350 + £175 = £525

    £3500 - £525 = £2975

24) 60% of the visitors are children.
    40% of the visitors are adults.
    The ratio of adults to children is 40:60 = 4:6 = 2:3

25) (a) $\vec{AB} = \vec{AC} + \vec{CB} = -a + b$

    (b) $\vec{CA} + \vec{AB} + \vec{BC} = a - a + b - b = 0$

26) The exterior angles of a polygon add up to 360°. The regular octagon has eight equal exterior angles so each exterior angle is $360° \div 8 = 45°$.
    Each interior angle is $180° - 45° = 135°$
    The interior angle of an equilateral triangle is 60°.
    $x = 135° - 60° = 75°$

27) (a) $(b+4)^2 = (b+4)(b+4) = b^2 + 4b + 4b + 16 = b^2 + 8b + 16$
    (b) $(c+7)(c-7) = c^2 - 7c + 7c - 49 = c^2 - 49$

28)

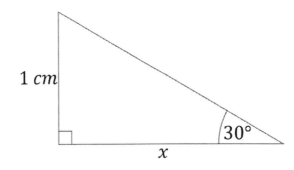

$\tan 30° = \frac{1}{x}$ and $\tan 30° = \frac{1}{\sqrt{3}}$ so $x = \sqrt{3}$

The area of the triangle is $\frac{\text{base} \times \text{height}}{2} = \frac{\sqrt{3} \times 1}{2} = \frac{\sqrt{3}}{2}$ cm²

29) (a) $x^2 - 2x - 15 = (x-5)(x+3)$

    (b) $x^2 - 2x - 15 = 0 \Rightarrow (x-5)(x+3) = 0 \Rightarrow x = 5 \text{ or } x = -3$

30) The area is the area of a rectangle minus the area of two semi-circles.
    $= 2 \times 5 - \pi \times 1^2 = (10 - \pi)$ cm².

Paper 2 Answers

1) $3\frac{1}{2} \div 2\frac{3}{4} = 9\frac{5}{8}$

   On a Casio calculator, press the shift key followed by the fraction key to enter a mixed number.

2) You can enter each number on your calculator and convert it to a decimal.

   $\frac{3}{7} \approx 0.429$ $\quad$ $0.4 = 0.400$ $\quad$ $43\% = 0.430$ $\quad$ $\frac{21}{50} = 0.420$ $\quad$ 0.401

   Writing all the number with three decimal places makes the correct order easier to see.

   0.400   0.401   0.420   0.429   0.430

   The correct order is $\quad$ 0.4 $\quad$ 0.401 $\quad$ $\frac{21}{50}$ $\quad$ $\frac{3}{7}$ $\quad$ 43%

3) (0.45, 0.18)

4) 382.7 rounded correct to one significant figure is 400.

5) $2x(3x + 5) - 2x(x - 7) = 6x^2 + 10x - 2x^2 + 14x = 4x^2 + 24x$

6) $420 = 2^2 \times 3 \times 5 \times 7$ $\quad\quad$ $90 = 2 \times 3^2 \times 5$

   Use a factor tree or the FACT key on your calculator if it has one.

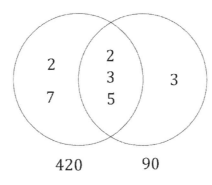

420 $\quad\quad$ 90

HCF = 2 x 3 x 5 = 30 $\quad\quad$ LCM = 420 x 3 = 1260 $\quad$ or $\quad$ LCM = 90 x 2 x 7 = 1260

7) 7   12   6   10   4   15   8   11

In order, the numbers are

4   6   7   8   10   11   12   15

The median is $\frac{8+10}{2} = 9$.

8) The probability that Bob is not late is 1 – 0.02 = 0.98

9) $15x^2y - 3xy^2 = 3xy(5x - y)$

10) $13.55 \leq x < 13.65$

11) (a) 2, 3, 5, 7, 11

(b) $1^3 = 1 \times 1 \times 1 = 1$ $2^3 = 2 \times 2 \times 2 = 8$ $3^3 = 3 \times 3 \times 3 = 27$

12)

$\frac{(7.23-12.49)^2}{\sqrt{2.32+4.656}} = 10.475343$

13)(a) $6.31 \times 10^5 = 631000$

(b) $(3.5 \times 10^8) \div (7 \times 10^2) = \frac{3.5}{7} \times 10^{8-2} = 0.5 \times 10^6 = 5 \times 10^5$

14) (a) The transformation is a rotation, 90° anti-clockwise, about the point (2,0).

(b)

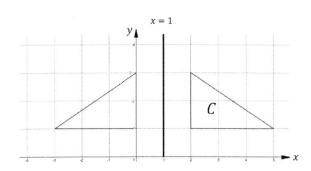

15) 12 stone 5 pounds = 12 x 14 + 5 pounds = 173 pounds

173 pounds = $\frac{173}{2.2}$ kg = 78.$\dot{6}\dot{3}$ kg

6 feet 2 inches = 6 x 12 + 2 inches = 74 inches

74 inches = 74 x 2.54 cm = 187.96 cm = 1.8796 m

Bob's BMI = $\frac{78.\dot{6}\dot{3}}{1.8796^2} = 22.3$ correct to one decimal place.

16) £74.75 = 115% of the original price.

The original price is $\frac{£74.75}{115\%} = £65$ You could also calculate $\frac{£74.75}{1.15}$

17)

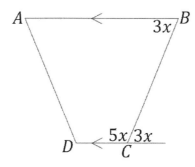

The angles marked $3x$ are corresponding angles, so they are equal.

The angles, $5x$ and $3x$ at C are adjacent angles on a straight line so they add up to 180°.

$5x + 3x = 180°$

$8x = 180°$

$x = \left(\frac{180}{8}\right)°$

$x = 22.5°$

18) (a) The modal class is the class with the highest frequency.

    The Modal class is $1.8 \leq h < 1.9$

(b)

Height, h (metres)	Frequency	
$1.5 \leq h < 1.6$	3	3 x 1.55 = 4.65
$1.6 \leq h < 1.7$	5	5 x 1.65 = 8.25
$1.7 \leq h < 1.8$	10	10 x 1.75 = 17.5
$1.8 \leq h < 1.9$	12	12 x 1.85 = 22.2
$1.9 \leq h < 2.0$	2	2 x 1.95 = 3.9
Totals	32	56.5

If the total of the frequencies is not given in the question, then you need to add the frequencies.

The estimate of the mean is $\frac{56.6}{32} = 1.77$ m

19) The sixth term is $3 \times 6^2 - 2 = 106$

20) 21 33 42 19 32 25
     46 37 28 33 35 42

You might first like to put the numbers into an unordered stem and leaf diagram by filling in the numbers, column by column,

```
1 |    9
2 |    1    8    5
3 |    3    7    3    2    5
4 |    6    2    2
```

and then sort them.

```
1 |    9
2 |    1    5    8
3 |    2    3    3    5    7
4 |    2    2    6
```

Key 2|1 = 21

The median speed is 33 km/h.

21)

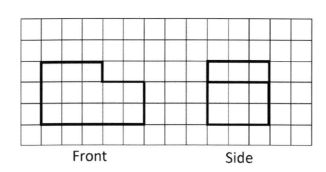

22)

155 pounds is about 70 kg.

23) $v = u + at$

Subtract u from both sides.

$v - u = at$

Divide both sides by a.

$\frac{v-u}{a} = t$

$t = \frac{v-u}{a}$

24) Ann receives 7 − 4 = 3 more shares than Bob.
Three shares = £24.
One share = £24 ÷ 3 = £8.
Ann receives 7 x £8 = £56.

25) By Pythagoras' theorem $AC^2 + 3^2 = 5^2$ or $AC^2 = 5^2 - 3^2$
$AC^2 = 25 - 9 = 16$
$AC = \sqrt{16}$
$AC = 4$ cm

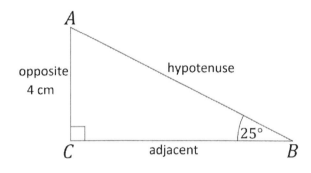

$\tan 25° = \frac{4}{BC}$
Multiplying both sides by BC
$BC \tan 25° = 4$
Dividing both sides by $\tan 25°$
$BC = \frac{4}{\tan 25°} \approx 8.6$ cm

You can also find the answer using a formula triangle.

Cover up the "Adjacent" and you see $\frac{\text{opposite}}{\tan(\text{angle})}$

26) The area of the wall is 6 x 2.5 = 15 m².
Bob will need 15 ÷ 5 = 3 litres of paint.
Bob needs to buy 2 tins of paint.
The cost will be 2 x £4.75 = £9.50

27)

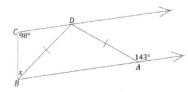

Angle $DAB = 180° - 143° = 37°$ (Angles on a straight line add up to 180°)
Angle $DBA = 37°$ (Base angles of an isosceles triangle are equal)
Angle $CBA = 180° - 98° = 82°$ (Allied angles add up to 180°)
x = angle CBA – angle DBA
$x = 82° - 37° = 45°$

28) The capacity of the container is the area of the circular base multiplied by the height
Capacity = $\pi \times 15^2 \times 40 = 28274$ cm³.
1 litre = 1000 cm³.
Capacity = 28.27 litres.
28.27-15=13.27.
The container can hold another 13.27 litres.

Paper 3 Answers

1) (a) If you enter $\frac{11}{12}$ or $11 \div 12$ on your calculator it will say one of the following:

   $\frac{11}{12}$ $\qquad$ $0.91\dot{6}$ $\qquad$ 0.9166666667

   If the answer is in the first or second form convert it to the third form and round to three decimal places. The answer is 0.917

   (b) Enter $\frac{3}{5} \times 120$. The answer is 72.

2) (a) Write all the numbers with three decimal places.
   0.301 $\qquad$ 0.300 $\qquad$ 0.030 $\qquad$ 0.031 $\qquad$ 0.310 $\qquad$ 0.003

   Now put them in order

   0.003 $\qquad$ 0.03 $\qquad$ 0.031 $\qquad$ 0.3 $\qquad$ 0.301 $\qquad$ 0.31

   (b) 0.03 x 0.3 = 0.009. The numbers with a product of 0.009 are 0.03 and 0.3.

3) (a) $\frac{3}{5}$ are girls, $\frac{2}{5}$ are boys. The ratio of girls to boys is 3:2

   (b)

Male	Female	Total
5	7	12
		36

   5 + 7 = 12
   36 ÷ 12 = 3
   7 x 3 = 21
   There are 21 female teachers.

4) $x \rightarrow \div 5 \rightarrow \times 4 \rightarrow 28$
   $x \leftarrow \times 5 \leftarrow \div 4 \leftarrow 28$

   $28 \div 4 = 7$
   $7 \times 5 = 35$
   Bob thought of 35.

5) Bob bought $50x + 20y$ books.

   If you find this type of question difficult, imagine Bob bought some particular numbers of small and large packets.
   For example, if Bob bought 3 large packets and 5 small packets then he bought $3 \times 50 + 5 \times 20$ packets. Now replace the numbers with the letters.
   $x \times 50 + y \times 20 = 50x + 20y$ packets.

6) If Amy gets 7 times as much as Bob then ratio of their shares is 7:1
   7 + 1 = 8
   £96 ÷ 8 = £12
   7 x £12 = £84
   Amy receives £84.

7) (a) 1249   (b) 1150

8) £191880 is the original price of the house + 23% of the original price of the house.
   That's 123% of the original price of the house.
   $$\frac{£191880}{123\%} = £156000$$
   You can also calculate this as $\frac{£191880}{1.23} = £156000$

9) (a) The mean $= \frac{23+29+21+27+20+29+22+19}{8} = \frac{190}{8} = 23.75$
   (b) The range = largest – smallest = 29 – 19 = 10
   (c) The mode is the most common value. The mode is 29.

10) $v = 5 + 6t$
    (a) When $t = 7$, $v = 5 + 6 \times 7 = 47$.
    (b) Subtract 5 from both sides
        $v - 5 = 6t$
        Divide both sides by 6
        $\frac{v-5}{6} = t$

    $t = \frac{v-5}{6}$

11) The 5 must be in the units place so that the number is odd.
    There are then 3 choices for the 10s place, then 2 choices for the 100s place and 1 'choice'
    for the thousands place. 3 x 2 x 1 = 6
    Bob can make 6 odd numbers with these cards.

12) $\frac{1}{20}$ of 240 = $\frac{1}{20} \times 240 = 12$

    15% of 240 = 15% × 240 = 36

    240 − 12 − 36 = 192

    192 students got grade 6 or 7.

    3 + 5 = 8   192 ÷ 8 = 24

    5 × 24 = 120
    120 students got grade 6.

13) The shape is a triangular prism.
It has 9 edges, 6 vertices and 5 faces.

14) 10% of £280 = £28.
£280 − £28 = £252.
£252 ÷ 12 = £21
Each payment was £21.

15) $5x - 4y = 5$ (1)
$5x + 2y = -7$ (2)

Equation (2) − equation (1) gives
$2y - -4y = -7 - 5$
$2y + 4y = -12$
$6y = -12$
$y = -\frac{12}{6}$
$y = -2$

Substituting into equation (2) gives
$5x + 2 \times (-2) = -7$
$5x - 4 = -7$
Adding 4 to both sides
$5x = -7 + 4$
$5x = -3$
$x = -\frac{3}{5}$ or $x = -0.6$ if you prefer a decimal. Both answers are correct.

16) $y = \frac{1}{2}x^2 - x$

If your calculator has a table mode, you can use this to obtain the table of values. Ask your teacher or tutor if you don't know how to do this.

You can also work out the answers one by one. $\frac{1}{2} \times (-2)^2 - (-2) = 4$

x	-2	-1	0	1	2	3	4
y	4	1.5	0	-0.5	0	1.5	4

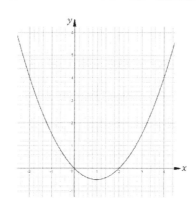

17) $ab = -12$. You need to find two integers with a product of -12.

You can ignore the minus sign at first.
$1 \times 12 = 12$
$2 \times 6 = 12$
$3 \times 4 = 12$
Because the product is negative, one of the numbers must be positive and the other must be negative.
$a + b = 4$. Choose 6 and 2, the pair that have a difference of 4.
$a = 6, b = -2$

18) (a) The cross section is made up of two rectangles.

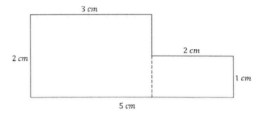

The areas of the rectangles are $2 \times 3 = 6$ cm² and $2 \times 1 = 2$ cm².
The cross sectional area is $6 + 2 = 8$ cm².
The volume of a prism is cross-sectional area x length.
The volume of this prism is $8 \times 2 = 16$ cm³.

(b) The total surface area is the sum of the areas of all the faces.

There are six rectangles and two L shapes.

The total area of the rectangles is

$5 \times 2 + 1 \times 2 + 2 \times 2 + 1 \times 2 + 3 \times 2 + 2 \times 2 = 28$ cm².

(This is the perimeter of the L shape multiplied by 2)

The area of the two L shaped faces is $2 \times 8 = 16$ cm².

The total surface area is $16 + 28 = 44$ cm².

19)
(a) $6x - 2 < 2x + 7$
Subtract $2x$ from both sides
$4x - 2 < 7$
Add 2 to both sides.
$4x < 9$
Divide both sides by 4.
$x < \frac{9}{4}$ or $x < 2.25$ if you prefer the decimal.

(b) $-3 \leq x < 2$

20) time = $\frac{\text{distance}}{\text{average speed}}$

The time required to drive 20 miles at 40 mph is $\frac{20}{40} = \frac{1}{2}$ an hour.

The time required to drive 50 miles at 60 mph is $\frac{50}{60} = \frac{5}{6}$ of an hour.

The total time is $\frac{1}{2} + \frac{5}{6} = \frac{4}{3} = 1.\dot{3}$ hours.   Don't ignore the dot over the 3.

average speed = $\frac{\text{total distance}}{\text{total time}}$

Bob's average speed for the whole journey is $\frac{70}{1.\dot{3}} = 52.5$ mph.

21) Will the corner hit the ceiling when the bookcase is stood up.

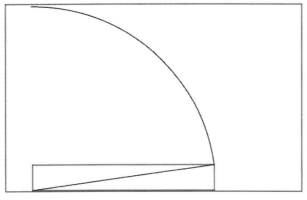

If the length of the diagonal is d then $d^2 = 28^2 + 202^2$.
$d^2 = 41588$ and $d = \sqrt{41588} = 203.93$ cm.
It will be possible to stand up the bookcase.

22) $1.3 \times 10^7 \times 3.6 \times 10^{105} = 1.3 \times 3.6 \times 10^{105+7} = 4.68 \times 10^{112}$.

23) $(2x + 3)(3x - 4) = 6x^2 - 8x + 9x - 12 = 6x^2 + x - 12$

24) $BC^2 = 4^2 + 5^2 = 41$ so $BC = \sqrt{41}$
$ED^2 = \left(\sqrt{41}\right)^2 - 5^2 = 41 - 25 = 16$ so $ED = 4$
The lengths of the sides of the two triangles are the same so the triangles area congruent by SSS.

25) It is not possible to calculate the probability as we do not know which cards are red and which are black.
If all the even numbered cards are red, then Bob the probability that Bob chooses an even numbered red card is $\frac{1}{2}$.
On the other hand, if all the even numbered cards are black, then it is impossible for Bob to choose an even numbered red card.

Printed in Poland
by Amazon Fulfillment
Poland Sp. z o.o., Wrocław